MARCH 2020

NATURE'S FREAK SHOW: UGLY BEASTS

THE UNSIGHTLY AYE-AYE

BY JANEY LEVY

Gareth Stevens
PUBLISHING

Please visit our website, www.garethstevens.com. For a free color catalog of all our high-quality books, call toll free 1-800-542-2595 or fax 1-877-542-2596.

Cataloging-in-Publication Data

Names: Levy, Janey.
Title: The unsightly aye-aye / Janey Levy.
Description: New York : Gareth Stevens Publishing, 2020. | Series: Nature's freak show: ugly beasts | Includes glossary and index.
Identifiers: ISBN 9781538246221 (pbk.) | ISBN 9781538246245 (library bound) | ISBN 9781538246238 (6 pack)
Subjects: LCSH: Lemur (Genus)--Juvenile literature.
Classification: LCC QL737.P95 L48 2020 | DDC 599.8'3--dc23

First Edition

Published in 2020 by
Gareth Stevens Publishing
111 East 14th Street, Suite 349
New York, NY 10003

Copyright © 2020 Gareth Stevens Publishing

Designer: Katelyn E. Reynolds
Editor: Monika Davies

Photo credits: Cover, p. 1 javarman/Shutterstock.com; cover, pp. 1-24 (curtain background) Africa Studio/Shutterstock.com; cover, pp. 1-24 (wood sign) Rawpixel.com/Shutterstock.com; cover, pp. 1-24 (marquee signs) iunewind/Shutterstock.com; pp. 5, 17, 21 David Haring/DUPC/Oxford Scientific/Getty Images; p. 7 Anna Veselova/Shutterstock.com; p. 9 Visuals Unlimited, Inc./Thomas Marent/Getty Images; p. 11 Mark Carwardine/Photolibrary/Getty Images; p. 13 IUCN Red List of Threatened Species(http://www.iucnredlist.org), species assessors and the authors of the spatial data (http://www.iucnredlist.org/technical-documents/spatial-data)/Chermundy/Wikipedia.org; p. 15 (main) Ariadne Van Zandbergen/Lonely Planet Images/Getty Images; p. 15 (inset) STEPHANE DE SAKUTIN/AFP/Getty Images; p. 19 Rob Cousins/Bristol Zoo via Getty Images.

All rights reserved. No part of this book may be reproduced in any form without permission in writing from the publisher, except by a reviewer.

Printed in the United States of America

Some of the images in this book illustrate individuals who are models. The depictions do not imply actual situations or events.

CPSIA compliance information: Batch #CW20GS: For further information contact Gareth Stevens, New York, New York at 1-800-542-2595.

CONTENTS

Say Hi to the Aye-Aye 4
Strange Appearance 6
Surprising Adaptations 8
Finger Food 10
Aye-Ayes' Island Home 12
Life at the Top 14
Mates and Mothers 16
Sounds and Scents 18
Endangered 20
Glossary 22
For More Information 23
Index 24

Words in the glossary appear in bold type the first time they are used in the text.

SAY HI TO THE AYE-AYE

Do you know what a primate is? It's any animal from the group that includes humans, apes, and monkeys. The aye-aye (EYE-EYE) is a primate, but it's hard to tell the aye-aye is a primate based on its looks. In fact, this primate looks so strange it was first believed to be a rodent, like a mouse or rat!

The aye-aye's appearance and the way it lives make it one of the most unusual primates around. Inside this book, you'll learn all about this surprising animal.

STRANGE BUT TRUE!
Aye-ayes have a rodent-like face and a tail like a squirrel, which is another familiar rodent.

THE AYE-AYE'S SCIENTIFIC NAME IS *DAUBENTONIA MADAGASCARIENSIS*. IT'S A LOT EASIER TO SAY AYE-AYE!

STRANGE APPEARANCE

If you look at an aye-aye, you'll instantly see how strange it looks. The aye-aye's pale, rodent-like face stands out against its dark body and long, bushy tail. Huge black ears rise above its pale face, and the aye-aye's bright, yellowish-orange eyes shine. The aye-aye's front teeth grow continuously, just like those of a rodent.

Aye-ayes also have large, strange-looking hands. Each finger has a long, curved claw. Their third finger is extra-long and so thin it looks like it's just skin and bones!

STRANGE BUT TRUE!
Aye-ayes have big toes that act like thumbs. Using their big toes, they're able to hold onto branches with their feet and hang upside down.

AYE-AYES ARE THE ONLY PRIMATES WITH TEETH THAT GROW CONTINUOUSLY.

7

SURPRISING ADAPTATIONS

Aye-ayes don't look unusual just to look unusual. Their strange features are **adaptations** for their way of life. Aye-ayes are active at night. Their dark fur serves as **camouflage** so they can move around without being spotted.

Aye-ayes have big eyes which help them see in the dark. They also have a special **layer** at the back of their eyes. This special layer is called the tapetum lucidum, and it reflects, or gives back, light. The tapetum lucidum is what gives aye-ayes their excellent night vision, or eyesight.

STRANGE BUT TRUE!
Aye-ayes use their front teeth to gnaw bark and their hard food. This gnawing action wears down their teeth. If their teeth didn't keep growing, they'd wear away!

THE TAPETUM LUCIDUM IS WHAT MAKES THE AYE-AYE'S EYES GLOW IN THE DARK. DOGS AND CATS HAVE THE SAME LAYER IN THEIR EYES.

FINGER FOOD

Aye-ayes eat mostly bug larvae. They use their extra-long middle finger, teeth, and big ears to find and eat larvae.

Aye-ayes use their middle finger to tap quickly on tree branches where the larvae live. When their big ears hear a hollow sound, this means there's a larva tunnel inside. They then use their teeth to bite a hole in the branch. Next, the aye-ayes stick their third finger into the tunnel and pull out the larva. And, they have a tasty treat!

STRANGE BUT TRUE! Aye-ayes also eat fruits, nectar, nuts, and seeds. They use their teeth to break open nuts and hard fruit coverings. They then use their middle finger to dig out the insides.

THE AYE-AYE'S BIG HANDS HELP IT HOLD SECURELY ONTO A TREE BRANCH WITH ONE HAND WHILE IT HUNTS FOR FOOD WITH THE OTHER HAND.

11

AYE AYES' ISLAND HOME

If you want to see aye-ayes in the wild, you'll likely have to travel to the other side of the world. They're only found on the island of Madagascar, which is off the southeastern coast of Africa. Madagascar is about 250 miles (400 km) from the African **continent**. The island's animals, including aye-ayes, are quite different from the animals of Africa.

Many aye-ayes live in the **rain forests** along the eastern coast of Madagascar. They're also found in other areas of the island.

STRANGE BUT TRUE!
Aye-ayes belong to a group of primates called lemurs. Madagascar has over 100 kinds of lemurs and 95 percent are in danger of dying out. Lemurs are only found on Madagascar.

WHERE TO FIND AYE-AYES

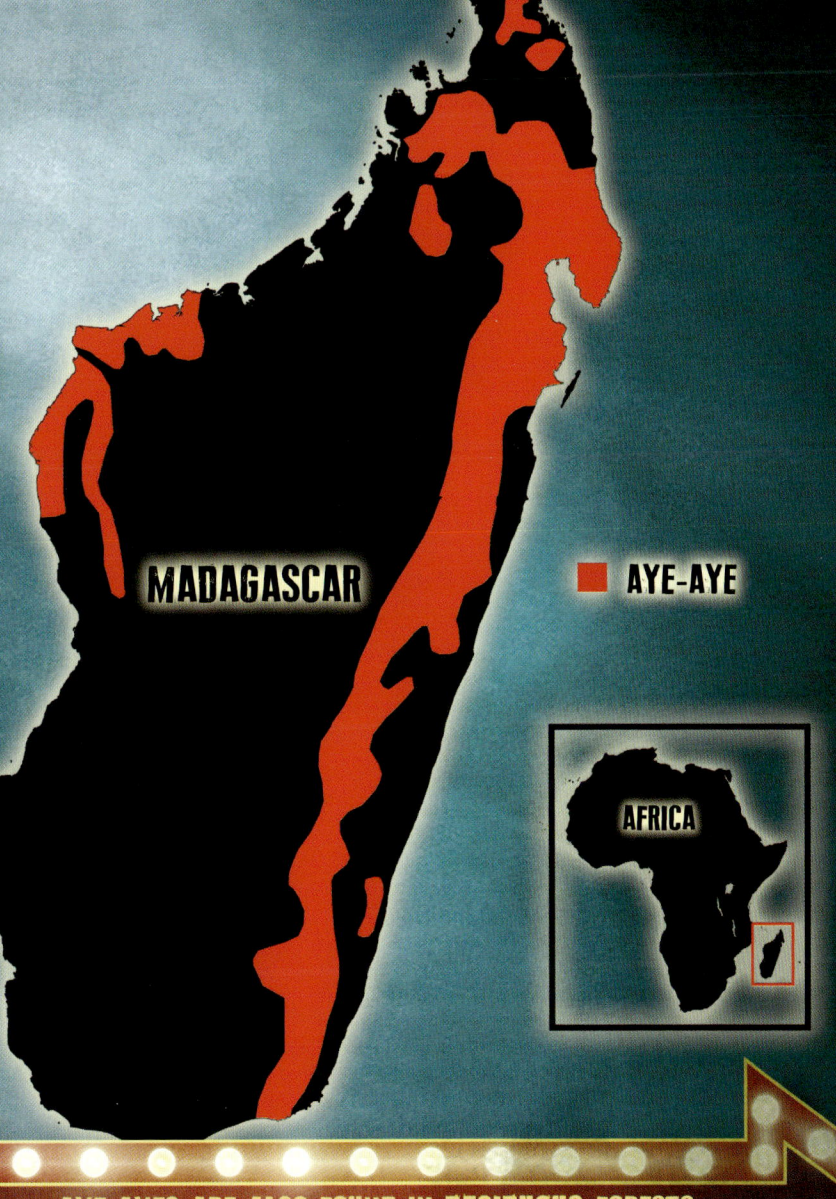

MADAGASCAR

■ AYE-AYE

AFRICA

AYE-AYES ARE ALSO FOUND IN DECIDUOUS FORESTS ALONG MADAGASCAR'S NORTHWESTERN COAST.

LIFE AT THE TOP

Aye-ayes spend most of their life high in trees, moving between the leafy branches that form the trees' **canopy**. They build ball-like nests in the forks of branches. This is where they sleep during the day. Their nests are made of leaves and small branches and have a single opening. Each nest is usually home to just one aye-aye.

Aye-ayes are **solitary** animals that hardly spend time together. They usually only meet up to **mate**. Female aye-ayes raise their young, but otherwise stay alone too.

STRANGE BUT TRUE! Aye-ayes eat, sleep, travel, and mate in the trees. It's fairly uncommon to see an aye-aye out of a tree, though they have been spotted on the ground!

FOSSA

LIVING HIGH IN TREES HELPS KEEP AYE-AYES SAFE FROM PREDATORS. THEIR MAIN PREDATOR IS THE FOSSA, A POWERFUL, CATLIKE ANIMAL.

MATES AND MOTHERS

Female aye-ayes are in charge when it comes to mating and raising young. When a female is ready to mate, she makes calls to draw males to her. She usually mates with more than one male.

After mating, aye-ayes return to their solitary lives. The female has a single baby about 5 to 6 months later. She raises her baby alone. She nurses, or gives milk to, her baby for about 7 months, and the baby stays with her for up to 2 years.

STRANGE BUT TRUE! Baby aye-ayes start to play with their mother when they're around 2 months old. They jump, chase, and run along branches.

MOTHER

BABY

THE MOTHER AYE-AYE KEEPS HER NEWBORN BABY IN THE SAFETY OF THE NEST WHILE SHE HUNTS FOR FOOD.

SOUNDS AND SCENTS

Although aye-ayes are mostly solitary animals, they do have ways to **communicate** with each other. One way they communicate is through sounds. Baby aye-ayes call to their mother when they're apart. Aye-ayes also make calls when there's danger or they're fighting.

In addition, aye-ayes communicate by marking their territory with their scent. They may rub their cheek, chest, or rear end along an object to leave their scent on it. Sometimes—like dogs—they pee on an object to mark it. Ew!

STRANGE BUT TRUE!
A baby aye-aye separated from its mother might make a "creee" call.

AN AYE-AYE MAKES A "HAI-HAI" SOUND WHEN THEY RUN FROM DANGER. THIS MIGHT BE HOW THEY GOT THEIR NAME!

ENDANGERED

Sadly, aye-ayes are one of the most **endangered** species, or groups, of primates. As forests are cut down, aye-ayes have suffered a loss of **habitat**. Losing their habitat—and the food it provides—has caused aye-ayes to steal crops from farms. Farmers kill aye-ayes to stop them from eating the crops.

In addition, some people kill aye-ayes because they believe the creatures are signs of evil. Some think an aye-aye's arrival will bring bad luck to a village. Luckily, today's laws help keep aye-ayes safe.

STRANGE BUT TRUE! It was once believed aye-ayes had become extinct, or ceased to exist. But they were rediscovered in 1957.

ZOOS AND OTHER PLACES ARE RAISING AYE-AYES TO MAKE SURE THESE SPECIAL CREATURES DON'T DISAPPEAR FOREVER.

GLOSSARY

adaptation: a change in a type of animal that makes it better able to live in its surroundings

camouflage: colors or shapes in animals that allow them to blend with their surroundings

canopy: the upper branches of a forest

communicate: to share ideas and feelings through sounds and motions

continent: one of Earth's seven great landmasses

deciduous: having trees with leaves that fall off every year

endangered: in danger of dying out

habitat: the natural place where an animal or plant lives

layer: one thickness of something lying over or under another

mate: to come together to make babies

rain forest: a tropical forest that has very tall trees and receives a large amount of rain

solitary: tending to live or spend time alone

FOR MORE INFORMATION

BOOKS

Arnold, Quinn M. *Aye-Ayes*. Mankato, MN: Creative Education, 2019.

Bluemel Oldfield, Dawn. *Aye-Aye*. New York, NY: Bearport Publishing, 2018.

Owings, Lisa. *Aye-Aye*. Minneapolis, MN: Bellwether Media, 2014.

WEBSITES

Aye-Aye
kids.nationalgeographic.com/animals/aye-aye
Learn about aye-ayes and see some photos on this website.

Aye-Aye
www.denverzoo.org/animals/aye-aye
Discover more about aye-ayes here.

Lemur
animals.sandiegozoo.org/animals/lemur
Are you interested in learning more about different species of lemurs? Find information about this interesting group of animals here.

Publisher's note to educators and parents: Our editors have carefully reviewed these websites to ensure that they are suitable for students. Many websites change frequently, however, and we cannot guarantee that a site's future contents will continue to meet our high standards of quality and educational value. Be advised that students should be closely supervised whenever they access the Internet.

INDEX

baby 16, 17, 18, 19
bark 8
branches 6, 10, 14, 16
claw 6
ears 6, 10
eyes 6, 8, 9
face 4, 6
farmers 20
females 14, 16
fighting 18
finger 6, 10
food 8, 10, 11, 19, 20
fossa 15
fruits 10
habitat 20
hands 6, 11
larvae 10
laws 20

lemurs 12
Madagascar 12, 13
males 16
mothers 16, 17, 18, 19
nests 14, 19
night 8
play 16
predators 15
primate 4, 7, 12, 20
scent 18
scientific name 5
tail 4, 6
tapetum lucidum 8, 9
teeth 6, 7, 8, 10
territory 18
toes 6
trees 14, 15
zoos 21